TECHO EN MEXICO
THE MEXICAN ROOF
2009 —

edition: 'ʌŋgewʌndtə

Book Series of the University of Applied Arts Vienna
Edited by Gerald Bast, Rector

SpringerWienNewYork

REVISITED

Springer Wien New York

CONTENTS

WOLF D.
PRIX

"The Mexican Roof:

It is only through experience that information becomes knowledge."

———

BÄRBEL MÜLLER

The relationship of aesthetics and sustainability is a subject of powerful relevance. The deep interest in exploring this relationship in architecture was one of the initial motives for starting a design process, an adventure and discourse, whose outcome is the roof in Mexico. Seven years after the idea was born to let students create a gathering place for a young institution in Oaxaca, and five years after completion of its physical manifestation in the form of a bamboo roof, reflections and discussions on it and its broader meaning culminated in revisiting the project and a conference in Oaxaca at the end of October 2009.

From the very beginning of the project the rhetoric of sustainability, which was common at the time, had been replaced by the concept of sustainable behavior of architectural form, materiality and environmental performance. This generated the design process and served as the main aesthetic, but also programmatic device. To evaluate its sustainable performance in the literal sense of its "capacity to endure" was the real interest in revisiting the roof. How would the bamboo have aged? How would people have been appropriating the space? How has the project influenced its context spatially and socio-culturally? How would one perceive the roof as an object of architecture, now, five years later, after having built it with one's own hands in an earlier period of one's life?

Especially with regard to projects carried out in the non-Western world by students from the Western world, people rarely get the chance to honestly reflect on "what have I been doing here?" on site years later. The opportunity to discuss it with a group of incredibly experienced and highly committed Mexican, Austrian and American architects was also a once-in-a-lifetime encounter. The present publication should be seen as an attempt to share an insight into this experience; and as an attempt to eternalize the project, as we don't know what tomorrow will bring.

———

TECHO EN MEXICO
THE MEXICAN ROOF
PROJEC

TECHO EN MEXICO – 96° 13´ W 16° 33´ N

Seven students from Studio Prix built a community centre in Oaxaca, Mexico (2002-2004)

A group of seven students from the studio of Professor Wolf D. Prix have been working on the design and realization of a community centre in the Province of Oaxaca, Mexico. The task was to create an architectural signal and landmark for the Instituto Tonantzin Tlalli – an organization promoting sustainable agriculture – as impetus for future developments as well as to research and apply new building technologies. Over six intensive months on site the project group and an assistant professor undertook the building process themselves, with the help of local labour. Techo en Mexico is architectural signal and community space.

6.5 kilometers of bamboo were woven into a two-layer grid, which serves as the primary structure of a free-formed roof. This roof creates a dialogue with the lines of the surrounding landscape. It shelters the new community from the sun, provides shade and collects rainwater for watering the plantations of the Instituto Tonantzin Tlalli. The project is a successful example of student realizations and a contribution to the discourse on sustainable and formally sophisticated architecture in the context of non-Western cultures.

Design and realization: **TERCER PISO ARQUITECTOS**
Jean Pierre Bolivar Martinez, Dominik Brandis, Alexander Matl, Giulio Polita, Florian Schafschetzy, Rüdiger Suppin, Rupert Zallmann
Project management and supervision: **Bärbel Müller**
Design and concept assistance: **Bärbel Müller, Reiner Zettl**
Construction and detail assistance: **Franz Sam**
Structural assistance: **Klaus Bollinger/Bollinger+Grohmann, Franz Sam**
Client: **Instituto Tonantzin Tlalli, César López Negrete (Director)**
Co-workers: **Andrea Börner, Pedro Cortés, Félix González, Martin Hess, Luis Juárez**

Location: **Paraje Bonanza, Ejutla de Crespo, Oaxaca, Mexico**
Building type: **Community building**
Surface area: **250 m²**
Completion: **May 2004**

96° 13´ W 16° 33´ N
Paraje Bonanza
Oaxaca Mexico

PROJECT CHRONOLOGY

November 2002	César López Negrete, founder of the Instituto Tonantzin Tlalli (ITT) located in the Mexican province Oaxaca, presents the ambitious 10-hectare permaculture project at Studio Prix in Vienna and invites students to design its future community centre.
March – July 2003	A group of seven selected students from Studio Prix research, conceptualize and design a community space for ITT as a semester program in Vienna. Seven individual projects are developed and then synthesized into a single project from May 2003 onward. Roof and tower are chosen as the given typological background.
April 2003	On a research excursion to Oaxaca (by Bärbel Müller, Jean Pierre Bolivar Martinez, Rüdiger Suppin) the exact project site – located on the only high ground of the ITT terrain – is defined and thoroughly documented. Research is carried out on potential building materials, and bamboo evolves to be the most interesting material to investigate and design with. Local working and living conditions are examined so as to prepare the future project phase on site.
June – July 2003	Models and plans of the current stage of the design are exhibited at the AzW as part of the exhibition "Just Build It – 10 Years of the AzW" (curated by Bärbel Müller, Reiner Zettl). The final review of the semester takes place in the exhibition space.
July 2003	The form-finding process is specified with a physical hanging model to gain knowledge regarding the structural behavior of the free-shaped roof (supported by Klaus Bollinger). Also, TERCER PISO is founded so as to have a name and as a means of appearing as a "legal entity".
July – September 2003	The design of the roof is developed in more detail and a structural model is built in the scale of 1:10. A 1:1 workshop takes place in Herzogenburg, Lower Austria, to test a variety of building technologies and to simulate a prototypical double-shell bamboo structure with wooden roof battens (organized and supervised by Franz Sam).
October 2003	The group of seven students and the project manager (Bärbel Müller) arrive in Paraje Bonanza, Ejutla de Crespo, where they work, build and live for the following six months.

November 2003	A sculptural outdoor structure, which serves as kitchen and office, is built with split bamboo to test its behavior, limits and potentials, as well as to acclimatize the team to local working conditions. Parallel to this, an exact 1:10 model is built, informed by the new findings of the 1:1 experience. Also, the footprints of the roof and tower are transferred onto the site with a 1:1 outline, repositioning both according to their actual topographical context. At the end of the month the group travels to Vera Cruz, on the Caribbean Coast 500 km to the east, to visit a bamboo farm where different species are examined, selected and ordered.
December 2003	Without machines, but with basic tools only, the group works on the foundations: digging, reinforcing and mixing concrete by hand with the help of local mason Pedro Cortés. On December 18 the long-awaited truck from Vera Cruz – packed with bamboo – arrives on site.
January 2004	At the beginning of the new year construction with bamboo can begin. Six-meter-long bamboo poles are split and tied with wire to create lengths up to 25 meters. The lengths are bound on the floor to create two flat nets that later become the two layers of the roof structure. January 21, both bamboo grids are raised and propped up with the aid of local hands and different kinds of scaffoldings. The roof is held in place by scaffolding, poles and ropes for the following months. Its silhouette is visible from far away.
February 2004	Wolf D. Prix travels to Oaxaca to review his students' work and to discuss upcoming design decisions on site. Over the course of February the two layers of the roof are split apart. To fix them into shape, 320 spacers of differing length are produced. These spacers create tension between the two grids and stiffen the roof, with the distance between the two layers varying between 65 cm and zero from the center to the edges.
March 2004	Back to concrete works: to channel the stress on the pivot-jointed steel column into the roof, ferrocement crosses are introduced wherever the supports meet a spacer. Solid concrete foundations are built, serving as ballast against the wind load. Dimensioning, positioning and detailing are based on an intensive four-day workshop on site (supervised by Franz Sam).

April 2004	April 6, the final presentation under the roof takes place with Wolf D. Prix and special guests Raimund Abraham, Gerald Bast, Klaus Bolllinger, Ernst Maczek-Mateovics and Karolin Schmidbaur. The opening of the solo exhibition "El Centro Comunitario" is celebrated together the day before at Casa Barragan in Mexico City (curated by Bärbel Müller, Andrea Börner). Central to this exhibition is the 1:10 physical model made in situ. Back to work in Oaxaca, the roof is covered with galvanized sheet metal. The floor area under the roof is paved with bricks, defining a continuous surface between the exposed foundation blocks of the steel columns. Parallel to this, the solar tower, which had no foundations before, is formalized and finalized: a solid water tank is built to collect the rainwater harvested from the roofs, and also serving as the tower's foundation.
Mai 2004	After several farewell fiestas, all project participants leave Paraje Bonanza. A second exhibition is set up and opened at the UNAM architectural faculty in Mexico City (curated by Bärbel Müller).
Mai – September 2004	The project is exhibited as part of the group exhibition "Rock over Barock" at Kunsthaus Muerz, Styria (curated by Reiner Zettl). Later this exhibition will also travel to Berlin, Aedes East.
November 2004	The Tercer Piso group wins an appreciation prize at the 2004 Awards for Experimental Trends in Architecture.
March 2005	"Das Dach (in Mexiko) 96° 13´ W 16° 33´ N" at AzW is a solo exhibition on the project that shows its making as well as the final architectural object (curated by Bärbel Müller, Rüdiger Suppin). The 1:10 physical model has been transported from Mexico for this event. *Prinz Eisenbeton 5: Thecho en Mexico. / The Mexican Roof. 96° 13´ W 16° 33´ N* is published by Springer as a detailed documentation of the project.
September – November 2006	The project is exhibited at the 10th International Exhibition of Architecture, La Biennale di Venezia, as part of the exhibition "Rock over Barock" at the Magazzine del Sale (curated by Reiner Zettl).
November 2009	The Mexican Roof Revisited, "Sustainability versus Aesthethics" Conference in Oaxaca

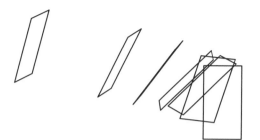

solar panels
optimized positions toward the sun

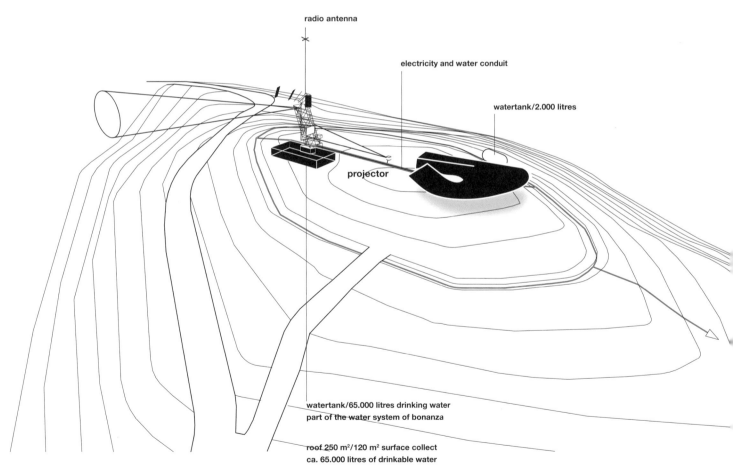

radio antenna

electricity and water conduit

watertank/2.000 litres

projector

watertank/65.000 litres drinking water
part of the water system of bonanza

roof 250 m²/120 m² surface collect
ca. 65.000 litres of drinkable water
per year (500 mm/m²)

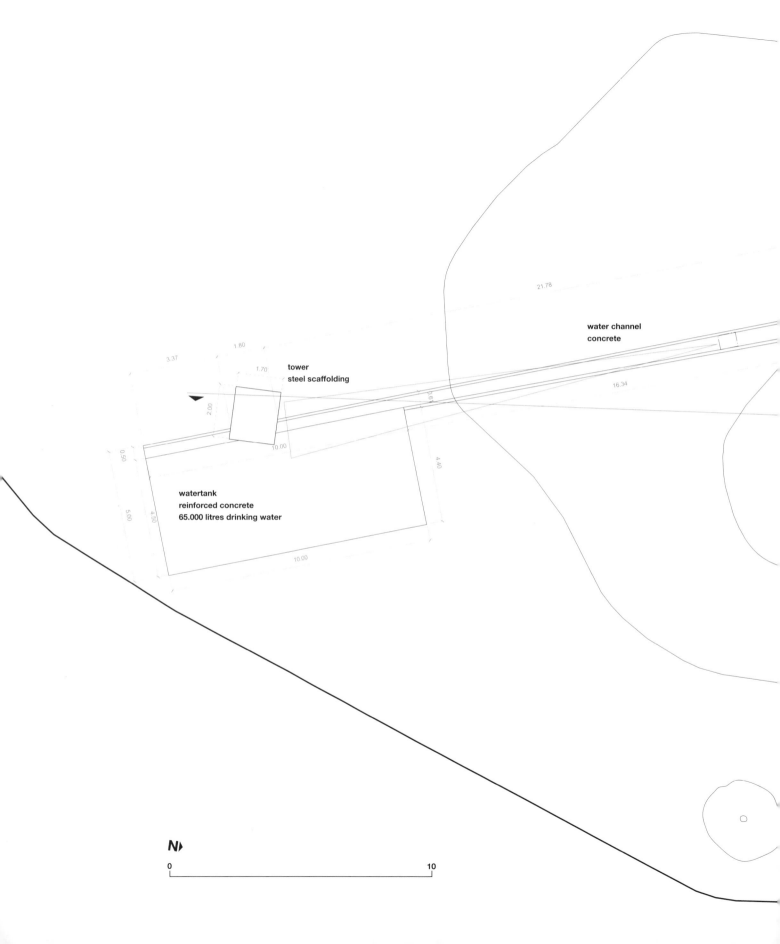

water channel
concrete

21.78

tower
steel scaffolding

1.80

3.37

1.70

2.00

16.34

0.50

10.00

4.40

watertank
reinforced concrete
65.000 litres drinking water

5.00

4.50

10.00

N▸

0   10

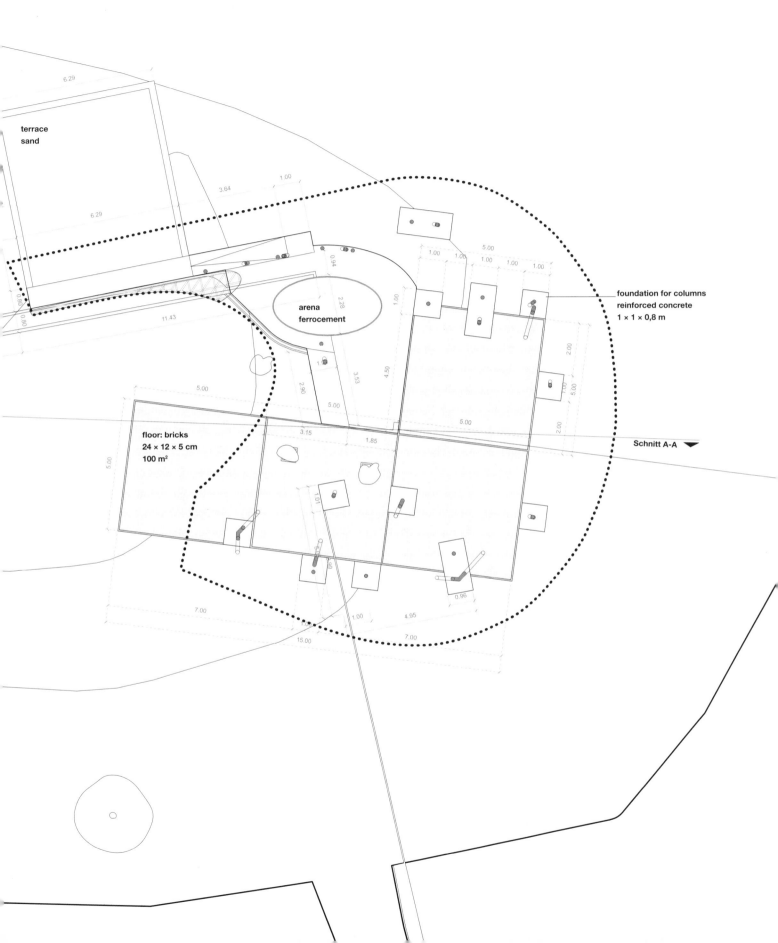

terrace
sand

6.29

6.29

3.64

1.00

0.94

arena
ferrocement

2.28

11.43

1.00

1.00

5.00

1.00

1.00

1.00

1.00

1.00

foundation for columns
reinforced concrete
1 × 1 × 0,8 m

2.00

2.00

5.00

5.00

4.50

3.53

2.90

5.00

5.00

floor: bricks
24 × 12 × 5 cm
100 m²

5.00

3.15

1.85

5.00

Schnitt A-A ▼

1.01

0.96

7.00

1.00

4.95

15.00

7.00

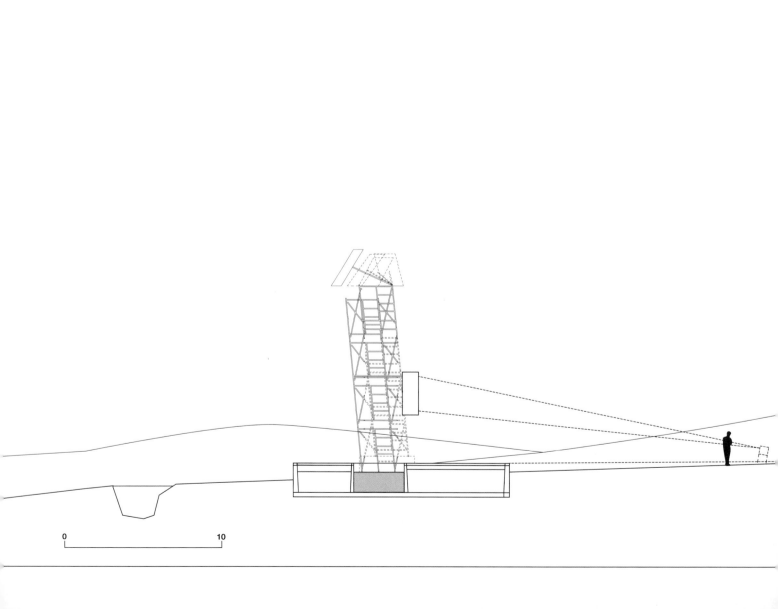

←
Ejutla de Crespo
6 km

0   10

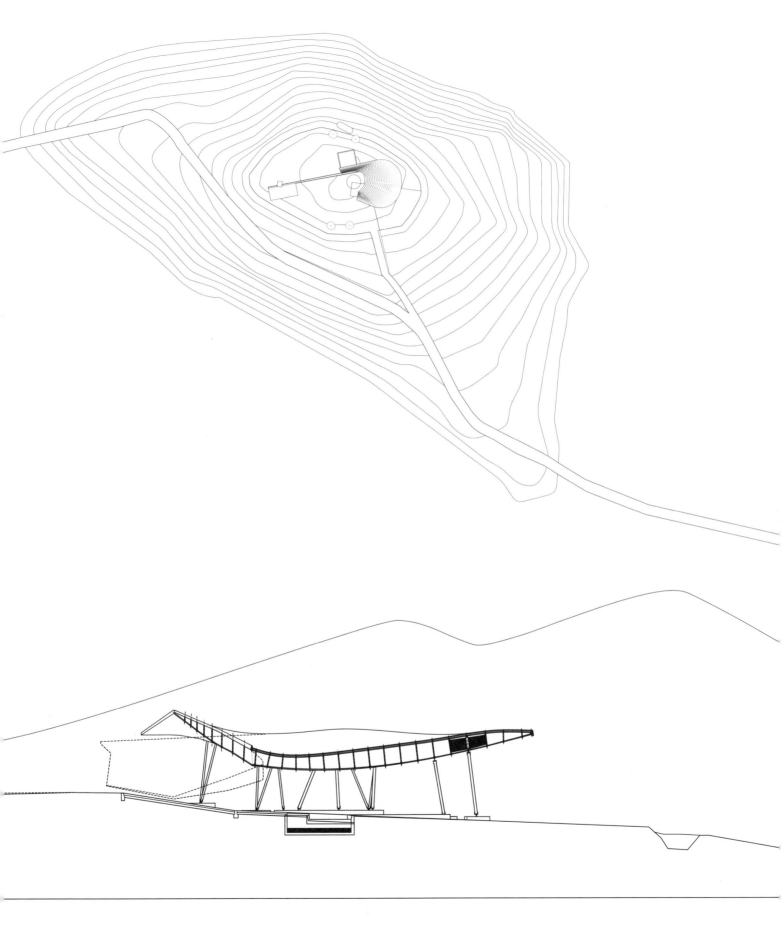

TECHO EN MEXICO
THE MEXICAN ROOF
GALLERY

REVISITED

The Roof in perfect shape in November 2009, five years after its completion

Roof, solar tower and a mud house that was built later

Reinforcing ferro-cement cross which channels the stress on the supports/steel columns into the roof

Florian Schafschetzy, Rupert Zallmann on the roof

Rüdiger Suppin, Andrea Börner on the solar tower

Florian Schafschetzy

Félix González

Rupert Zallmann

Reiner Zettl, Peter Strasser

Carl Pruscha, Hannes Stiefel

Klaus Bollinger, Franz Sam and Bärbel Müller

Solid concrete foundations function as ballast against the wind load

The reinforced arm of the roof

The roof seen from the west, surrounded today by a vegetable garden

César López Negrete, Wolf D. Prix

Celebration under the roof

César López Negrete presenting Tercer Piso and their local co-workers

Raimund Abraham, Michael Rotondi, Carl Pruscha

Día de los Muertos - Day of the Dead celebrated from 31st of Oct. to 2nd of Nov. Each building gets its own Ofrenda (altar)

Rupert Zallmann, Giulio Polita

Jean Pierre Bolivar, Giulio Polita, Reiner Zettl, Bärbel Müller, Andrea Börner

Jean Pierre Bolivar: reunion with the local project team

Franz Sam

TECHO EN MEXICO
THE MEXICAN ROOF
STATEME

Between Now and Then

RAIMUND ABRAHAM
Architect
USA
Taught at SCI-Arc Los Angeles, Cooper Union, New York

"The memories which lie within us are not carved in stone; not only do they tend to become erased, as the years go by, but often they change" Primo Levi

On the road to Ejutla, the familiar route leading to the town of Miachutlan and over the Sierra Madre to my house in Mazunte, coming from Oaxaca my eyes were searching for the electrical power plant on the left-hand side of the road, which I remembered as a signal to slow down and turn off the road to follow the tracks towards the hills of the barren Mexican landscape, anticipating the first glimpse of the memorable "bamboo-cloud" I had left, unfinished, five years ago.

The sudden and unsuspected barrier of a highway, radically dissecting the landscape and disrupting our path, turned all memory in a fraction of a second into the reality of "now". Only after an odyssey of underpasses and turns were we once again on our way, following the windblown, dusty tracks leading us towards the building we were revisiting.

"Isn't memory", I asked myself "a diversion of the imagination in the quest to recall what is irretrievably lost?" But then I told myself that it also allows me to confront, measure and critique the reality of "now" with the imaginary recollection of "then".

Where I thought I remembered the position of the structure within the landscape of rolling hills, a most unsettling and alien object appeared. Triangular in shape and disturbingly bright, it was unrecognizable at first, but as we approached it revealed its banal origins: a gabled roof of an inconspicuous and sordid building, void of any indigenous character and authenticity.

And behind it, threatened by the other building's mass, but defiant like an endangered species fighting for survival, stood the edifice, we had come to celebrate, the "Wolkenkuckucksheim", as I called it, a city built by birds in the sky.

I remember it standing alone, hovering like a cloud over this magnificent landscape, fragile but sturdy and radiating a call for a new architecture, an experimental architecture in an ancient land, ancient and revolutionary. But looking at the development of the site now, it was a cry into the void.

Whatever the architectonic consequences are, the "sky-roof" becoming a "tin-roof", the "bamboo-cloud" becoming profane as a result of utility, the spirit of this edifice will endure and the monumental effort by its builders should be celebrated forever. ▬▬▬

La Tortilla

MARGARITA BARNEY
*Director of Instituto Tonantzin Tlalli (ITT)
Mexico*

I must say that for Grupedsac and for the Insituto Tonantzin Tlalli to have the opportunity to receive students from Vienna and have them, under the direction of a world famous architect, build a roof using a natural renewable resource for a useful purpose was certainly a privilege. I envision an ITT that could become a showcase of architectural solutions. Not only for the poor, our target group, but for everyone.

The ITT is a place for permaculture in action. A place to train, teach, and offer environmentally friendly solutions for the poor enabling them to become self sufficient with regard to water, food, shelter and energy. But also to become more productive, to learn how to protect the fertility of their land and nature. We want our small farmers to regain their dignity and to take charge of their own development.

Can Vienna offer the ITT a home design that people could learn to build, which is socially, economically and environmentally viable and yet beautiful?

Grupedsac is deeply grateful for La Tortilla (a popular name given by people attending workshops) which has become a symbol. ▬▬▬

Applied socio-cultural responsibility: The Mexican Roof

DR. GERALD BAST
*President of the University of Applied Arts Vienna
Austria*

Architecture consists of more than pure form imbued with function. Being one of the most visible manifestations of all applied arts, architecture requires social responsibility, energy and resource efficiency, as well as a permanent drive for innovation and progress.

The Mexican Roof can be perceived as a powerful manifestation of this new form of considerate architecture. Serving as an assembly place and water-collector – for permacultural purposes – through its roofing, it goes far beyond providing simple shelter. Instead it symbolizes social interaction, use of local materials and the decentralized provision of basic needs. Finally, the project not only shows the willingness to combine form and function with ecologically sound construction methods, but also demonstrates the transformative power of theory and ideas put into action.

It is the drive towards action and social change through art, science and culture, the will to make a difference, which Prof. Prix's students realized in a powerful architectural statement in Oaxaca. Trying to make a difference is not only crucial for all of us, it is also what the University of Applied Arts has been founded on; something we try to live up to every single day. ———

JEAN PIERRE BOLIVAR MARTINEZ
Architect
Austria
Alumni of the Studio Prix at the University of Applied Arts Vienna

On our last visit to Mexico with Tercer Piso Arquitectos, it was fascinating for us to notice that the German pronunciation of the Spanish word "correcto" sounded quite different from its Spanish pronunciation, adding an undertone that would ironically suggest the opposite of its actual meaning. Due to the phonetic similarity between "correcto" and "korrekt", this difference became clearly noticeable, especially for Spanish native speakers.

The construction of THE MEXICAN ROOF clearly marks the transition between academic and professional life. Now, five years later, being surrounded by great colleagues who always were a great source of inspiration and for whom I feel the deepest admiration, I can make the following statements about THE ROOF and architecture itself:

THE MEXICAN ROOF is a correct building. KorreKTo.

With the advent of the digital age and the implementation of new tools in both design and construction processes, common sense has become a rather obsolete tool. KorreKTo.

Beauty is the most powerful force mankind has ever invented. KorreKTo.

Usually, architecture comprises five minutes of creativity and five years of patience in which one puts everything else aside hoping to get five minutes of creativity at least once a day. KorreKTo.

Architectural concepts are becoming more and more powerful. They have to, especially when they lack convincing design.

Intellectual colonialism doesn't exist, it cannot exist, but slavery can – and is voluntary. KorreKTo.

Architecture needs great investments, not great sacrifices. KorreKTo.

In architecture, authenticity creates style: it becomes recognizable, desirable and above all reproducible. KorreKTo.

What makes architecture stronger sometimes makes it more vulnerable at the same time: Not having been seen before. KorreKTo.

The next paradigm in architecture will be: find new solutions to deal with forces that in most cases pull in opposite directions: politics, ecology, economy, market requirements, sustainability, fashion... KorreKTo.

———

KLAUS
BOLLINGER
*Structural
Engineer
Germany
Professor at
the University
of Applied
Arts Vienna*

The Roof of Ejutla – Five Years later

One part of the project was the creation of a roof structure, providing a shaded space that was protected from rainfall, collecting the rainwater for an underground reservoir.

The generated space was to form a landmark and at the same time to be integrated into the landscape so that it appeared to be a floating, weightless sculpture.

On visiting the construction site shortly before completion five years ago, it was evident that both criteria had been fully satisfied by the chosen design, but also by the materials used for the roof construction. An organic, functioning roof shelter had been created, which thanks to its floating appearance could be easily adapted to the surrounding landscape. A truly unique landmark. Yet it was not foreseeable whether the natural material of bamboo would last in the local environment and whether the site itself would be accepted as a meeting place.

This was reason enough to check up on it five years after first visiting the site. "Techo de Tortilla", as it is almost affectionately called by the local population, still stands. Heat, rain and wind could not damage the construction. With pride, the builders could declare that the construction as well as the materials used have proved their worth.
No improvements were required and no serious repairs were necessary. But also the dream for the place to develop into a sort of meeting site for the farmers of the region has really come true. The Instituto Tonantzin Tlalli (ITT) has meanwhile constructed further buildings, laid out around the roof, serving as accommodation for Mexican and international guests visiting to participate in workshops and seminars about new ecological technologies.

For me, the roof of Ejutla is an example for the fact that an innovative, aesthetically sophisticated structure with sustainable material can be done. ———

ANDREA
BÖRNER
Architect
Austria
Lecturer at the
University of
Applied Arts
Vienna

Nearby or at a Distance

The first thing that now flashes through my mind when I remember my arrival at the construction site in Bonanza is the architectural model of the roof. There it was, planned to assume a strategic position in the territory; highly exposed for the later gathering of rainwater to irrigate the adjacent grounds and to communicate with the local surroundings. An object completed with great accuracy and detail, ten times smaller than the ambitious goal, but still very real, already able to initiate a dialogue with the landscape, and the further aim of contributing to international discourses.

Six years ago I arrived in Oaxaca after a two-week journey on a sailboat across the Atlantic with its endless horizons, slowly untying my regular patterns of daily behavior. The rhythm of the journey seemed to perfectly blend into the daily rhythm around the construction site. Living and working in a breathtaking landscape, at 1,600 meters above sea level in close vicinity to the people I knew, with locals and volunteers from other places involved. We inhabited the territory, temporarily occupying niches and corners in-between the well laid-out tubes of the irrigation system, which follow the contour lines of the sloping hillside.

During the building process the area of focus and investigation expanded from this small hill into the wider surroundings, and occasionally drew all kinds of people together – helping hands from the nearby town Ejutla, German students from another construction site at the outskirts of Oaxaca, travelling carpenters, architectural students of UNAM and visitors to the Barragán House where the project has been exhibited in the meantime. Translating the model into its architectural reality seemed promising on all levels. After mounting the last rivets, literally only a few seconds before the first heavy rains, we reverted to earlier routine, leaving the construction site in a hurry to catch our flights home to Austria, which we had booked a long time ago.

Six years later after a 24-hour trip, heading from airport to airport, the site seemed to be more distant and much more difficult to reach, though already known and signposted in the surrounding area. Bonanza's dry ground had turned into a green landscape occluding all traces and marks of our stay. On the contrary, the roof still retains traces of the construction phase; small numbered plates and auxiliary constructions, which labeled the single elements for coordination and communication. Maybe for this reason the roof so closely resembles a scaled-up version of the earlier model. Still at an exposed position it may renounce its strategic potential for the sake of the experiment; an object tested in different sizes of reality. Despite its perfect intertwining with the landscape, it retreats from the resulting proximity of further development, not being able to play out the spatial potential as the strategic intervention of sharing the hilltop with other structures, even though it visibly represents a small section of a greater idea. However, as a model it still bears all the potential and promises it could fulfill.

Today the tin roof is clearly visible in satellite images accessible through the web marking the spot on the globe, and the related narratives are fully unfolding in the multiplicity of spoken and unspoken interpretations. ▬▬▬

DAMIAN FIGUERAS MIN SON Architects Mexico

"...all architectural schools should be closed for five years" Raimund Abraham exclaimed on the way to the site at Paraje Bonanza, Ejutla, Oaxaca, a veritable no man's land.
As architectural design takes advantage of the new digital technologies there is a loss in the way materiality can inspire a project. Computer sketching is becoming a way of designing but by its very nature it detaches form from materiality. Due to the speed of evolution this thought might have a nostalgic dimension but something certainly not to be forgotten is that the ultimate expression of architecture is a physical intervention. The Mexican Roof is an interesting example of how architectural education bridges these changes.

Six months, seven students, local material and one goal: to reproduce a design that was created over ten thousand kilometers away in an extremely different climate by people who don't speak the language. That was not only facing a complex construction process but the birth of something new. Isn't it difficult enough to build something in your own city? An icon was created through its function. So it might not be about closing down schools but of widening their scope.

The conference "Sustainability vs. Aesthetics" was held to revisit this iconic project which the locals had dubbed the Tortilla Roof. Leading architects such as Raimund Abraham, Rozana Montiel, Wolf D. Prix, Carl Pruscha, Mauricio Rocha, and Michael Rotondi were the constellation of people discussing the subject. A strong concluding statement was that as the misuse of the term sustainability increases its meaning while depleting its strength. Opposition to the s-word was manifested by the white cotton costumes, panama hats, cigars and all-road cars. "Nothing is preserved if we intervene in it", Raimund Abraham said while puffing on his cigar.

But the Mexican Roof has to be understood as a catalyst. Here is not the place to be judge its beauty or efficiency, over which one could argue, but to assess its consequences and implications. It is a way of understanding current issues through a different teaching methodology that might serve as a guide for instruction. Let's hope for continuity. ▬▬▬

CÉSAR LÓPEZ NEGRETE
Architect
Mexico

The Instituto Tonantzin Tlalli based in Oaxaca works in an area with high unemployment, massive migration, vast areas of unused, eroding land, and a great need for development alternatives. Our aim is to help people help themselves by offering a learning environment with examples of local food production, self-built housing, rainwater management, soil rehabilitation and renewable energy.

Our facilities were all built using local materials. We incorporated innovations in our buildings, such as rainwater storage, solar energy, ferrocement, or adobe reinforcement. Nevertheless, their design is deliberatefully conventional, as we wanted to respect local forms.

The construction of the Techo signified a break from this modus operandi. By partnering with Studio Prix from the University of Applied Arts Vienna, we intended to create a landmark that would help us draw more attention to our work, attention being a means of augmenting our impact, drawing a bigger audience, achieving more presence in the media, and influencing decisions by local and state officers. We got far more than we bargained for!

The Techo certainly contributed to increased visibility and prestige for our learning center. But the impact of this building affected the very nature of our organization. First, our local team had to learn to collaborate with an international group of young professionals, which also meant that we had to learn to communicate our strategic priorities to a different crowd. Our remote site was being visited by people from outer regions and faraway lands. We also hosted photographers, artists, and architects. The structure stirred up debate, which did not keep us from hosting all kinds of events in it: open courses, dance parties, crop harvestings, workshops … but most importantly, throughout this process, we learned to question many of our assumptions.

This building is loved and used constantly, and it is also a continuous source of debate. As a symbol, the Techo has a different meaning for different people, and this meaning is open to change over time.
For our organization, five years after its construction, the Techo signifies no less than a rebirth into a new stage in our work: more than ever, we are innovation-driven. We seek to combine ethics with aesthetics, we are open to change.... and we also believe that architecture can enable people to think the unthinkable. ▬▬▬

Somewhere nowhere

ALEXANDER MATL
Architect
Austria
Alumni of the Studio Prix at the University of Applied Arts Vienna

For me personally this project is a symbol of passion. A passion which in my opinion is not natural, but which we were allowed to learn. Looking back at that time it seems like an architectural moon landing with nothing else but an idea as baggage. An idea, which did not only contain the object but also the absolute will to leave an abstract study project and venture into a then unknown terrain – the terrain of reality. We have developed a self-conception which in spite of all adverse conditions recognizes obstacles as opportunities and improvisation as a method. The experiment developed into a premise. Frei Otto once wrote that bamboo should only be used dried and not in a split or pierced form – we took note of that and did the exact opposite. ▬▬▬

Techo Revisited

ROZANA MONTIEL
Architect
Mexico
Assistant Professor at the Ibero University in México City

"Story is ultimately about relationship. The soul of the land becomes the soul of our culture not through information or data alone, but through the metaphor and analogy of story." Peter Forbes

We arrived at the site like pilgrims to commemorate five years since the construction of the roof. As we gathered under it, we created a sense of community in resonance with the site.
It is through the story of the site and the collective that this project has become alive.

A global architecture – a roof with a view – and, in the local context – a view with a roof – establishes itself on the hill, becoming part of the scenery. The roof is now the witness to endless stories. ▬▬▬

BÄRBEL
MÜLLER
Architect
Austria
Senior Lecturer
at the University
of Applied
Arts Vienna

The changing same...

The process of initializing, conceptualizing, experimenting, designing, producing, living, representing, reflecting, memorizing and revisiting the TECHO project spans a timeframe of seven years, a route of thousands of kilometers between the major waypoint of its spiritual origin (Studio Prix 48° 12´ N 16° 23´ O) and that of its physical manifestation (Paraje Bonanza 96° 13´ W 16° 33´ N), as well as the creation of a global social network of individuals who have been involved in it in various specific ways.

Trying to characterize the fundamental nature of this project, the notion of REPETITION first comes to mind. Repetition in a temporal and spatial sense, regarding both event space and physical space, thereby the making and using as much as the physical appearance of this piece of architecture. Not identical repetition, or "sameness" but repetition as the changing same.

The attractiveness of the roof emanates from the very beauty of its woven bamboo net structure, which again derives from the principle of differentiated repetition: the bent kilometers of split bamboo along with the hundreds of bamboo distance-pieces of various heights, each member grown differently - as nature works – all create the impression of continuous variation between two (or more?) rhythms. "Repetition produces tension and clarification between rhythms while allowing for the emergence of dynamic and open structures"[1], no matter if we speak about polyphonic music or the kind of architecture described here.

Creating the bamboo structure was an extremely rhythmic process, which epitomizes the large scale and more male version of the basket weaving work of Oaxacan women, whose almost mechanical moving hands one can watch day by day right next to the entrance of the market in Oaxaca de la Ciudad. This repetition in everyday work and everyday life allowed for precision and concentration. There was also enough of surprise and improvisation to not fall into tedious monotony.

The appropriation of the space as a place to be and to gather already began during its construction phase and has happened ever since "in varying degrees of habitation, tranquility and landscape"[2]. In a cultural context which seems to nourish itself through repeated celebrations, the roof on the hill offers an ideal setting: a multitude of similar but not identical spatial situations, which again appear tremendously different according to the natural light conditions at a given moment, the specific programming of the space and the respective sequential speed in which it is experienced - watching, moving, dancing …. One of these numerous festive events was the recent visit of the roof. Being there, again, five years later, felt simultaneously like a loop, a unique moment in time, and at one point like a social emotional spatial experience beyond time.

"Order emanates from repetition…ritual orders both 'life' and 'art'.
Since each repeated event occupies a unique place in ontological time,
repetition subtends both stasis or consistency and dynamism."
Kofi Agawu ———

[1] Asgedon, Araya (2000). The Unsounded Space, in: White Papers, Black Marks – Architecture, Race, Culture, by Lesley Lokko, Athlone Press
[2] Lim, cj (2003). The First Garden, in: How green is your garden? by cj Lim, Wiley Academy

GIULIO
POLITA
Architect
Austria
Alumni of the
Studio Prix at
the University
of Applied
Arts Vienna

In one of his last interviews Ettore Sottsass Junior proposed, with his usual ironic attitude, a golden BIC ballpoint pen as a paradigmatic example of sustainable design. Ergonomic, generic, easily rechargeable, it would have been the answer to the widest range of different needs, avoiding by instinctive affection for the precious material the small ecological disaster of the daily throw-away.
Not the aristocratic Mont Blanc, just a BIC! Everybody has had to deal with a BIC at some time, from Alvaro Siza Viera to the secretary. Luxury for everybody and none of the pedantic ecological educational aims. Architecture is based on other social dynamics of identification: the desire invested in space defines its quality far more than the price of materials embodying the enclosure. Even more than its correspondence to the function representing the opportunity to reveal itself.
En fond du sac, l'emotion!

PS: The BIC Cristal is the most widely sold pen in the world and as of 2004 one hundred billion had been manufactured.

PPS: By the way, what could be the final result if we were to compare the ecological footprint of the most recent skyscraper by Sir Norman Foster to Saint Peter's Dome? ———

CARL
PRUSCHA
Architect
Austria
Taught at
Academy of
Fine Arts Vienna

Expectations were high, when we departed in different groups and vehicles from Oaxaca around midday to examine the object of our curiosity. Although we were all familiar with the work through publications and not least the wonderful essay about its creation by Raimund Abraham, experiencing a location yourself and the indulging in all its dimensions is always a new experience. Our journey with a fixed time of arrival led us through villages and landscapes into a world of undisturbed beauty. Occasionally we went along routes with renewed roads or even bridges and it felt like we were running away from these improvements and when leaving them often found ourselves driving into deserted country roads and thus exceeding our timetable by far. Finally we had already come very close to our final destination, when we suddenly found ourselves together with the other vehicles at the dead end of a field road from which our view fell upon a distinctive, tin-roofed large building on the opposite ridge of hills. Our object should also be located there, but we could not see it yet. After another return trip we finally arrived at the access road to our goal, which suddenly appeared before us. Unfortunately it was still partially hidden by steep tin roofing at the corners. It also seemed smaller than on the illustrations, but when we entered a feeling of guardedness and intimate closeness came over us, although we found ourselves under an into all directions open roof construction with wonderful views on the surrounding rolling hills.

After this emotionally so positive experience my reason started to work. But it was soon clear to me, that here irrational mathematics was at work, which called into question many seemingly unalterably fixed axiom's in natural science set by Western thinker's. These bamboo poles seemed to correspond to the archaic approach of ancient Indian cultures, whose order was more fixed on mythic beliefs than on the reason of our sober comprehension. Time was too short and the densely packed people under the roof too many to further ponder about it, but I wished to sit alone under the roof with the Indian medicine man like Castaneda had done before me, have a good herb in my pipe and philosophize with him about these poles.

When I went back out into the open and inspected the upper building, which had so unpleasantly caught our attention already from far away as well as a row of similar buildings, which as somebody told me, had been erected around the location, I came to realize that these, who ever may have built them, could never have come from the same Indian spirit that this mysterious structure seems to have sprung from. ———

MAURICIO
ROCHA
Architect
Mexico
Professor at
UNAM, Mexico
City. Ibero
University in
México City

Architecture is a craft. The craft of thinking. The craft of building thoughts. Academic speculation is essential. It is fundamental in daily practice and in the construction of buildings.

That an architecture student continues to experiment, using new digital technologies and perhaps reproducing a three-dimensional model in another scale, helps, but is never the same as the possibility of working in a real, physical scale – an exercise in constructing ideas – in a specific place with specific materials.

Experimenting and finding solutions, traveling and finding a different context from the habitual. Experiencing being and living in a place with other customs and another culture, isolated from the outside world and in close contact with the site in which one is intervening. These are parts of the fundamental process of making architecture – understanding that the process becomes a result and that the result becomes a metaphor of the process.

A roof that floats between the Oaxacan mountains opens paths to construct new ideas, new processes, and essentially new experiences that strengthen the task of making architecture. ━━━

Tale of two cities / Kyoto to Oaxaca

MICHAEL
ROTONDI
Architect
USA
Professor at
SCI-Arc Los
Angeles, Univer-
sity of Kentucky,
Arizona State
University

We were more than an hour south of Oaxaca city, in the rural areas. Exit signs were handmade and moveable. Direction of travel, geographical characteristics, driving duration, and memory served as our GPS. 'Exit here', someone said. The road from the highway was unpaved, winding up into the foothills towards our destination. We were not sure we were going in the right direction or if it was the correct road. It did not seem to matter. The road had been made by a mere scraping of the ground in the most efficient way, moving like a narrow multi-colored ribbon along the side of the hill. The varied landscape and vegetation distracted us from any concern we may have had of reaching our destination. The car in front of us kicked up enough dust to create a semi transparent veil which at times seemed to be a rainbow of light filtering the mountains in the distance. "Is that it" someone asked. "Is that the structure?"

"No, but it has a similar profile", he said.

My mind began to conjure images of Kyoto. I was there three weeks prior and visited several of the temple gardens in the foothills of the mountains that surround the northern boundaries of this city. Long approaches, overlay of vegetation and hills, perceived extensions and compressions of space and a calmness and coherence of experience was my recollection at that moment. Each time we recollect a prior experience, the present context transforms it, I once read. A superimposition of mountains were the equivalent of a superimposition of my visual thinking. There was an unfolding of my minds eye out into the landscape, triggered by this material layering. 'Layered landscape, layered perceptions, and layered thoughts, all compressing into the kind of flatness seen in a Rothko painting', I thought. The Japanese have a term for this, Shakkei – borrowing of distant scenery – background and foreground become one in the minds' eye. This type of embodied experience is 'centering'. It feels wonderful. Time stands still as long as you can hold it. "Is that it in the distance on that intermediate plateau?" we asked again. "Yes, there it is / was." He answered, as the road curved and roof structure moved out of view.

Soon we decided to go the remaining distance on foot. Walking along the inclined path was reminiscent of the uphill walk to Ronchamp, a long spiral procession. We had been sitting a long time viewing the landscape thru the windows of our car and now in it. The sound of the winds and the soil underfoot, the long shadows from the late fall afternoon sun. The colors of the natural landscape and suddenly a silver cloud with a profile like the rolling hills surrounding us. We were 'grounded' by the bands music. The 'silver cloud' was over a lot of people in a happy exchange, talking and listening and eating, and studying the structure of the cloud. It was made of bamboo, steel, concrete, and thin steel sheets. It was precisely what it needed to be to realize an aesthetic vision and to honor gravity. It's 'body language' spoke in silence of all this and more. If I listened with little to no interference I could 'hear' its story being told. It conveyed the history of its own creation and growth. Like a thousand year old tree, people were naturally attracted to the covered space it made underneath its immense canopy. A Communion of strangers.
It worked. It was as a gift to the place and the people. It was constructed cooperatively, like many 'barn raisings' before, to solve a practical need for a community space but with a longer vision of sustaining an essential aspect of our humanity – cooperative life.

Slowly approaching, passing under and through, around it entirely looking from ground to sky and then laterally along the continuous profile line of its eave superimposed on the hills in the middle distance and the mountain a distance beyond. One moving image of an object in space

and space defined by objects. "It was all one thing", I thought.
The roof enhanced and in turn was enhanced by the natural orders
processes of its surrounding landscape, at several scales.
Once again, Kyoto came to mind.
Procession, communal, invisible hand, resourcefulness, spontaneous
innovation, craft, Wabi-Sabi, and most of all Shakkei – borrowed
landscape, a perceptual device to compress of foreground, middle
ground, and background into one image and a conceptual device to
conjure the possibility of the infinite, and its twin, zero.
Although many of the problems the team of students and teachers
had to solve on **TECHO EN MEXICO** were practical and technical, what
resulted were inventions of the highest order, rising about the limits
of matter into the weightless realm of the poetic. ——

FRANZ
SAM
Architect
Austria
Senior Lecturer
at the University
of Applied Arts
Vienna

The wish to see the Techo again amounted to a faraway feeling of
possibility – just as then when everything was about keeping that
uplifted tortilla up there – and that it is still up there is not only a matter
of work and ability, but also of will. thus it is not the tornado that makes
roofs fly against gravity, but the power of will. ——

HANNES STIEFEL
Architect
Austria
Lecturer at the University of Applied Arts Vienna, Innsbruck and the University at Buffalo

El Reeuncuentro

If in today's search for other models of a future architecture one is able to witness a virtual dialogue between Gertrude Stein and Kazimir Malevich in, under and beyond a roof in the Sierra Madre del Sur in the south of Mexico, then this, indeed, is a significant experience.

A confrontation (even the first encounter) with great architecture is always a reunion, an unexpected reunion with a presence that was unintended by the authors. Such architecture is mostly characterized by cracks and obvious contradictions. Less obvious, at first, is the impact which needn't be understood immediately or completely – this is where real sustainability lies.

Since knowledge may only take effect when it is turned into experience, I like to see the roof as the spatial description of such an experience – one that generates further experiences; its impact will spread out more and in more ways than expected.

A great little piece of architecture! Hats off! ——

PETER STRASSER
Workshop Manager at Studio Prix
Austria
University of Applied Arts Vienna

"The Mexican Roof Revisited" brought together all project participants, the native population and renowned architects. They were invited to participate in a conference and a survey of the project, followed by an ensuing festival.

"The Mexican Roof Revisited" was an exciting encounter with a project that was concluded five years ago. And although it is not normal for students, it would be desirable that every student should be able to realize one of their own projects in a 1:1 scale during the course of their studies.

The Techo is in very good condition, because it was accepted by its users, which not only delighted our former students, who would maybe do some things differently today.

It is important that the university will also support such projects in the future. ——

RÜDIGER
SUPPIN
Architect
Austria
Alumni of the
Studio Prix at
the University
of Applied Arts
Vienna

About Felix and schools
The diversity makes the place richer
The mistakes make the storyteller smarter
The potentials never get lost
The dances everybody will remember

"Felix what do you think about the roof"
I think it is very good, it gave us a name.
It gave us a lot of festivals and events.
Especially a theater play by a German group in 2007
"el misterio más grande del mundo"
On this occasion a lot of school children came together"

I would like to introduce a young man and say what I know about him, which is not that much, but enough to bring you closer to a place called Bonanza. Let us call him "Felix de Bonanza", who I met the first time six years ago. Felix lives in southern Mexico in a small hut. Under his baseball cap he seems to hide a big thirst for any kind of technical or linguistic information available. This capacity got challenged as a community came from the north to this place called Bonanza with the goal of turning a piece of dry and eroded land into a self-sustaining ecosystem. Felix's new neighbor, Instituto Tonantzin Tlalli (ITT), came up with new tasks like building tree houses, windmills made out of old bicycles, tepees created with coffee bags and rudimentary radio stations to receive worldwide broadcasts.

A symbiosis between Felix and the community arose, because of their need for water irrigation systems and further technologies for renewable energies. Over the years international workshops on experimental building and agriculture turned this piece of land into a school based a certain idea of sustainability. For me the fascination of this place lies in the rudimentary aspects of learning and building with a direct visual impact on the environment.
Almost at the same time architecture students in Vienna were given the task of designing a community center with a focus on elementary architecture for the ITT in Bonanza, a place none of us had ever been to before. After a field trip to the Mexican site and a design phase back in the known world, we started to implement a specific language in this place in fall of 2003 and so a dialogue between two schools emerged. Looking back, the most intense part of this dialogue for me was the process of building itself.

Over six months an idea, the site and materials turned into a structure in a very direct way, which makes the construction readable with or without necessarily understanding the specificity.
The moment that something starts to become real (and depending on gravity) recent constraints do still exist but become transformed into something else. The process of giving way to something automatically results in a new project.
Now the built structures (roof and tower) provide shade, collect rainwater and carry a wind mill and solar panels.
In the end it seems less important if it was a play with intellectual syntax, mathematical grammar, political emotions or pragmatic reality that led to decisions in that building process. What counts is maintaining the potential for discussing all of them in a complete manner.
Is it that any kind of polarization neither helps to develop construction nor social environment but instead seems to be a motor for stagnation? Pursuing crazy ideas with (self-generating) energy, testing and experimenting as provocation counts as much as the ability to listen.

For me the most exciting moment in revisiting the project was when I heard that just recently, a few kilometers away, the next scaffolding tower with solar panels and a windmill had been erected. A small neighborhood has had electricity since then. And guess what, Felix was part of it. ━━━

RUPERT ZALLMANN
Architect
Austria
Alumni of the Studio Prix at the University of Applied Arts Vienna

Contrary to what some believed five years ago, the roof did not decay, was not blown away and was not dissembled by locals desperate for some building material. It settled in as a friendly alien, it continues to be maintained and the locals identify it as the "tortilla roof".
The surrounding community has learned from it, they have started to grow the same bamboo we had used for the roof structure and in the backyards they have erected copies of the energy tower we developed for the site.

The project 'Angewandte builds in Mexico' has turned out to be sustainable in its theoretical context, as this symposium and the way the locals are trying to make practical use of the roof have shown. Personally I am very grateful for having had the chance to develop a project beyond the confines of academia by finalizing an experiment in a 1:1 scale. I not only learned how to mix concrete but how to build a spaceship. ━━━

CARLA ZAREBSKA
Publisher
Mexico

I don't want to talk about architecture, because every single article that refers to this adventure is about architecture. About inclined planes and magnificent roofs designed by Wolf Prix, constructions that have inspired many others. About the muddy grounds of Oaxaca and about adobe walls, which Wolf isn't particularly fond of, being a Na'vi who likes to work with materials that bring him closer to the skies. Now it seems that I've started talking about architecture after all…

What I would like to talk about is the human in human beings. I would like to express how I felt when I saw Giulio, Rüdiger, Alexander, Jean Pierre, Florian and Rupert handing over their gifts to Pedro and Felix during the fifth anniversary celebrations of "The Roof"; how I felt when I noticed how much each and every one of these students had become attached to this place – a place that may not have anything in common with Vienna, their usual creative home, but nevertheless has become a part of them after six months of shifting, touching, walking, feeling and smelling the grounds of Oaxaca on which "The Roof" was to be built. I would like to believe that the country that is shaping the hearts of Pedro and Felix every day has also left a mark in theirs.

I also like to remember Bärbel with her red lips, always perfect, no matter what time of the day; to remember how she coordinated the work of her students from the Institute of Architecture at the University of Applied Arts Vienna with meticulous precision; to remember her as a grand lady from a faraway European country, with fair skin and golden hair resembling our beautiful grass fields and a woman whose brilliant instincts gave her absolute control of the situation.

The Day of the Dead celebrations in November 2009, with all of our Viennese friends present, seemed way too short. I could feel how the past six years had changed me and notice the empathy in Wolf, César, Bärbel and the students, who in the meantime have become architects too. I can still feel my joy and also theirs of having been part of such a unique encounter, an encounter that has given me lots of great memories.

Happy Anniversary! ——

*REINER
ZETTL
Art Historian
Austria
Assistant
Professor at
the University
of Applied Arts
Vienna*

5 Years Later

The roof is a success. As intended by the client it has created a place. Nobody had noticed the hill before and nobody would pay too much attention to the institution that now uses the place without its presence.

The hoisting of the roof is remembered as a collaborative finalization of the form. By means of its own weight and the positioning of the supports, the bamboo net that was laid out flat on the ground assumed its shape once it was pulled up. Neighbours participated in this special festive event and this has contributed to preserving its non-standard form.

It was surprising to see to which degree the materials assert themselves. Concrete foundations, steel columns and bamboo weave complement each other according to their respective tasks.
The roof seems more natural than the later adobe building next to it.
Now, more than at the time it was finished, one feels the spatial generosity of the dialogue with the topography unfolding in a cinematic framing of the panoramic view onto the surrounding hillside.
Carlo Scarpa meets John Lautner. ━━━━

TECHO EN MEXICO
THE MEXICAN ROOF
PANEL—

Hannes Stiefel

Bärbel Müller

Raimund Abraham

SUSTAINABILITY VERSUS AESTHETICS?

A conference hosted by the Institute of Architecture of the
University of Applied Arts Vienna
Participants: **Raimund Abraham, Rozana Montiel, Wolf D. Prix,
Carl Pruscha, Mauricio Rocha, Michael Rotondi**
Moderator: **César López Negrete**
Location: **Centro de las Artes de San Agustín, Etla. Oaxaca, Mexico**
Date: **Friday, October 30, 2009 at 4 pm**
Concept and Organization: **Bärbel Müller, César López Negrete**

PRESENTATION BY PANELISTS

*RAIMUND
ABRAHAM*

¡Buenas tardes! My name is Raimund Abraham. I'm an architect.
I live and work in New York. I first came to Mexico in 1966 and before
I came I prepared for my journey. I grew my moustache! When I went
back to America I shaved it off but suddenly I felt very lonely. So I decided to let it grow again and since then it has become part of Mexico.

The main reason why I'm here today is to celebrate the fifth anniversary of a unique edifice. When I first heard about it I was quite surprised
because I didn't know that one could have an anniversary of a building.
One usually celebrates the anniversary of a marriage. However, I left
here after the roof was completed and now I'm back again to celebrate
one of the most magnificent buildings ever made by students (or maybe
even in the history of architecture.)

I didn't bring my own work, because I probably anticipated the screen
was too small anyway. Instead, I will read a very brief manifesto,
"un chico manifesto", to clarify my views on sustainability and
aesthetics: While we all, who are meeting here today, enjoy the
hospitality of a magnificent country, a few thousand kilometers to the
north, the government of the United States is erecting a barrier of power
and fear at the border to Mexico. Merciless, oppressive, and undeniably
beautiful, this wall, built by the US Homeland Security Agency, meets
all criteria of sustainability and aesthetics. Thus, sustainability, void of
any ethical dimension, has become a catch phrase for all those who try
in vain to save the world by changing light bulbs and promoting green
buildings. I don't want to save the world but I shall always celebrate
and defend the human spirit. ¡Zapata Vive!———

Raimund Abraham

Rozana Montiel

ROZANA MONTIEL

Hello, my name is Rozana Montiel and I'm a Mexico City architect. I would like to begin my speech with a quote from Gregory Bateson: "There is an ecology of bad ideas, just as there is an ecology of weeds." It is said that 80% of the environmental impact of a product, a building or a system have its origins in the design phase. When put into practice, these bad ideas destabilize our ecosystem and therefore won't survive.

The metropolitan area of the Valley of Mexico has 22.7 million inhabitants, the Federal District has 8.7 million inhabitants and the State of Mexico is home to 14 million people. The question is how to approach the issues sustainability and aesthetics in one of the largest cities in the world? When we speak of sustainability we should not just get lost in a modern phrase that conveys a lot of enthusiasm but is void of any real meaning or content. Therefore, we have to be very careful when we use the term "sustainability".

The first project I am going to present is about urban recycling. The topic of recycling is of great personal interest to me and so I chose this example because it shows how recycling can be translated into architecture. What we do as architects is not just erecting new buildings. We also revitalize existing buildings that have been abandoned or disused. The group of buildings I'm presenting to you is located along lines 1, 2 and 3 of the Mexico City Metro and consists of 22 identical structures from the 1970s. Since the individual structures are all connected in some way, we could actually speak of them as one entity. Today, most of them are in very bad condition. They serve as taverns or have been abandoned completely. Possibilities for adaptation are manifold. However, every day, millions of people pass the area so our idea was to convert them into social spaces such as nurseries, gyms, cultural venues, areas of recreation and learning. We have already made a few drafts. The latest project proposal has been elaborated in cooperation with the Secretariat of Urban Development and Housing.
Here in the picture we see just one example of how the buildings could be used. In this case, the idea was to keep an extra storey which can be used freely.

With my next example I would like to address the question whether it is possible to approach the issue of sustainability in areas of extreme poverty. We have tried to answer this question using the municipality of Chimalhuacán as an example. Chimalhuacán, which is located east of Mexico City, in the State of Mexico, covers an area of approximately

Klaus Bollinger

Carl Pruscha

Wolf D. Prix

Raimund Abraham

7,000 hectares and has a population of 600,000. Our idea here was to consider sustainability as interaction, as something that has to do with new perspectives, with a systemic and holistic way of thinking.
In Chimalhuacán sustainability practically doesn't exist. The area has suffered an economic decline, 84% of the population has no access to education. The rubbish is dumped in open landfills, most sewers are open drains and the area is flooded on a regular basis.

So the challenge was to work with the people of Chimalhuacán in order to transform the place and create some kind of sustainability. This has to do with a systemic approach, with local economy, social transformation, generative logic, collective learning, ecological culture, ecosophy, evolutional change, urban recycling, fractal governance, integral synergetic self-empowerment, collective participation, cultural diversity, active ecology and auto-production. We applied sustainable strategies, worked with the community and realized a number of small-scale projects to improve the quality of life. One of these was our truck project.
I should maybe explain that in Chimalhuacán vegetation is very sparse because it is built on the ancient lake bed of Lake Texcoco and therefore has a high concentration of salts. So, the idea was to place this truck in front of the landfill, add some vegetation and see if people would finally make use of it, see what would happen to the place. In another exercise we placed the Virgin of Guadalupe at various waste deposits and observed people's reactions. What they did was they collected the figurines and built altars, thus changing the meaning of the place.
As I have already mentioned, sustainability is not simply about buildings and objects, it is also about social interaction.

After we had carried out these projects with the community the Government asked us to develop a concept for two eco-parks, a project we are currently working on in collaboration with students from the Ibero-American University in Mexico City. Some of them are also here today. Yesterday, Mauricio Rocha and I had a meeting with them at the University. The parks we are developing have a linear form.
One of them runs along the entire border and the other one on the high grounds of the "Cerro Chimalhuache", a recreation area for many locals.

One major issue there is the large amount of waste dumped on the hill. What we are planning to do is to build slopes and use the waste as building material. So, in this case, the problem itself holds the solution. Rather than seeking the perfect shape for its own sake we are trying to

Michael Rotondi

Carl Pruscha

solve problems, problems that arise from extreme poverty, and transform them into architecture. The point here is that we make architecture by using what is there and from that we create what I like to call ethical aesthetics. Thank you. ——

CARL PRUSCHA

I'm going to show you four small, rather modest projects and I will talk about them in a rather pragmatic way. So don't expect any philosophical deliberations on sustainability, which is the theme of this gathering. The term sustainability has always struck me as being one of those ugly, feel-good expressions. Since I never used the term sustainability myself this has been an excellent opportunity for me to think about it and reflect on what it actually means to me. When I looked it up in the dictionary I discovered that it meant "holding on to something" or "maintaining" or "carrying on things that are meaningful to the future." With this in mind I selected the following four projects because I think they fit in this category and qualify as something sustainable.

The first project is a small cluster of houses built in the vicinity of the Great Stupa of Bodanath in the Kathmandu Valley in Nepal. I lived there for about ten years in the 1960s and early 1970s. At that time I was in charge of drawing up a master plan for the conservation of the physical environment and cultural heritage of the valley on the basis of which UNESCO was to declare it a World Heritage Site. The Kathmandu Valley, a bowl-shaped area of about 20 by 25 kilometers, owes its development to the fact that it contains valuable alluvial soils from a prehistoric lake. The unique brick culture that emerged as a result of that is very rare on the Indian subcontinent. However, when Nepal became independent from India foreign influences began to take hold. Cement was introduced and the traditional brick architecture was in danger of being lost forever with historic brick structures being plastered with it. I thus considered it my responsibility not just to promote the conservation of old buildings but also to demonstrate how contemporary buildings could be constructed with traditional materials and construction techniques. We realized several model projects in which I employed the geometry of traditional forms like ancient mandalas in my architectural idiom. One example is a hostel I built for foreign researchers and artists who intended to spend some time in the valley.

The second project I will present to you is my own little lagoon bungalow which I built about ten years ago on the lush tropical shores of Sri

Raimund Abraham

Carl Pruscha

Michael Rotondi

Lanka overlooking the Indian Ocean. Before conceiving the design I looked for traditional examples of buildings in the area. The only thing I could find were a few tiny structures that fishermen had built for themselves as shelters while they were out in the water fishing. Residential buildings only reflected the colonial influence of more than four hundred years of foreign occupation and while they might have provided suitable housing for people in Portugal, Holland or England, their heavy solid walls and tiled roofs now stood in a climate of extensive monsoon rains and extreme humidity.

Finally I decided to build my bungalow on pilotis, elevated above the damp ground, and chose light-weight transpiring walls to guarantee proper cross ventilation. Since the wooden beams needed for such a construction were not available in the required dimensions, professional carpenters in the area were no longer able or willing to take on such an assignment. However, there were boys in the village who were able to do metal works and welding structures. So I decided to use telephone poles as the vertical structure and T beams for the horizontal structure. Smaller sized wooden beams were laid across the floors and ceilings. The walls were lightweight constructions, wooden frames filled with either glass elements or panels that were locally available. The building served its purpose very well for several years until the tsunami hit and it had to prove its sustainability. Since the floor was only about 2 meters above the ground and the ceiling including the roof was 3 to 4 meters higher than that, the main wave, which rose some four meters above the normal water level of the lagoon, crashed right through the building destroying partition walls as well as the furniture behind. The construction of the building, however, was not damaged at all while other buildings in the area were either entirely swept away or heavily impacted.

The third project that I would like to present to you was also built in Sri Lanka. It is a school building that originally stood near our bungalow and was completely destroyed by the tsunami. The new building was to be erected by local, mostly unskilled workers. I designed a reinforced concrete structure consisting of several cross-shaped columns and cassette-type roofs which would form individual rooms measuring about 6 x 6 meters. The local metal workers who had built my bungalow welded steel sheet forms that were used for the roof structure. They were temporarily supported by bamboo rods to be moved to the next section after the concrete was cast above. This method enabled us to minimize dimension of the concrete roof to about 2 ½ inches. Employing only a

Michael Rotondi

Carl Pruscha

minimum of the expensive material concrete we were able to achieve accurate forms and sufficient strength. On the inside, the classroom walls were detached from the structural columns and constructed as free-standing entities so that in the event of another tsunami they would collapse without endangering the main structure. This can be seen as another example of sustainable architecture for this island.

The final project I would like to show you is one I'm currently working on. It is related to habitats in an urban context using individual housing units. While the "one-family home" still is universally accepted as the standard form of housing, it has proven to be both ecologically and economically disadvantageous due to its enormous space require-ments and development costs. It is the epitome of anti-urban construc-tion. My photo shows a group of houses in the Tibetan town of Gyantse which is the only still existing example of a traditional housing type that has survived the Cultural Revolution in China and remained largely unchanged over the years. In this housing type the individual courtyard houses are a valuable element within the compact urban texture of the city. In order to provide for a new sustainable type of housing in subur-ban and decentralized areas I translated this typology into a European context where its compact arrangement of a large number of individual units can produce a living area of urban dimensions. ▬▬▬

MICHAEL
ROTONDI
The deepest genetic imprint of all living organisms is endurance. Sustainability normally stays with biological necessity and environ-mental necessity. Endurance also extends to our humanity, which is the most difficult, yet most profound aspect of our existence.
The projects that we'll be moving forward here are of various scales but they deal with everything from biological, environmental, but most importantly social, cultural and spiritual sustainability.
Houses that are made from 50% of recycled material and built in a very unconventional way; buildings that are more conventional in their sustainability.

And then, cultural sustainability is sustainability working with American Indians trying to help them keep their culture alive by listening to their stories and translating them into planning and architecture. Very tradi-tional systems are reading the sky and bring them to the earth.
The configuration of the buildings is a way for them to teach, to tell their stories by looking at the buildings.

Hannes Stiefel

Bärbel Müller

Mauricio Rocha

And then there is economic sustainability. There is no construction industry on these reservations, so we developed a construction industry starting by teaching the students and then developing more businesses. This is my homie, he travels with me.
And then there is education which is one of the most important things that need to be sustained. This is a school of architecture in Texas.
And then there is my own sustainability, which is one of my teachers.

―――――

MAURICIO ROCHA

Good evening. Well, I also think that the term sustainability bears a lot of risks and can seem a bit odd when it is used in the wrong context. Aesthetics is of course always a friend to someone who works with art. Here in Mexico we benefit from the fact that every single day we are confronted with problems. For architecture, the best way to confront these problems is to deal with them and much of this implies the question of how to achieve dignity in space, dignity and culture, and create better architecture with limited financial means.

As regards ephemeral architecture, I had the opportunity to create pieces of art. The project I am referring to is a museum located in the center of the city. Its slanted structure was equipped with a floor whose function was to solve the problem with the inclination. The museum director decided to remove the existing floor so I quantified its individual pieces and used them to build a room inside the room, thus creating a new spatial experience. The installation lasted for two months and cost no more than 500 Mexican pesos (50 dollars) plus the effort of the people working with us. With a very limited capital we managed to transform the energy of a place and create something totally new.

Another example is a project I realized in Iztapalapa, a district of Mexico City and an area of great poverty. I was supposed to build a school for blind people and was presented with a terrain of 14,000 square meters full of illegally deposited gravel and waste material from the nearby houses. We did not have the means to dispose of it, so we did some landscaping and transformed 70% of the land by building a series of slopes that would function as enclosure for the school building. Afterwards we added different fragrant plants creating a garden of odors for the blind people. By generating a new microclimate we man-aged to convert an existing problem into a new and positive experience.

César López Negrete

My next example is a project, which I had the pleasure of presenting to my colleagues. I am talking about the Oaxaca School of Plastic Arts where I had the opportunity to use an enormous amount of soil that was left over from another construction project. The material was to be removed so I asked whether I could use it for my project. With this material I built a sort of crater that would define a new topography for the school building inside, creating a new atmosphere, a new place, a new garden the university didn't have before. For the walls inside we used another type of soil. Based on a logical economy we can explore new architectural structures which are becoming more and more readily available to us.

Finally, we have come to the most urban project in Mexico City. Mexico was once full of rivers and creeks, there was a lot of water but today, almost nothing of it is left. We are now confronted with a lost forest and a contaminated river that is almost dead. In this context we were given the opportunity to realize a project that will last for many years and change this place for the better. We started the project with a survey to help us understand the characteristics of informal architecture. This is crucial because this kind of architecture evolves from inside the culture, from civil anarchy where we can find a lot of positive and interesting aspects. When these buildings begin to consume our forests we must not abandon them. Instead, we have to understand them and use them to create a type of urban acupuncture that will transform the place into something new.

I this context we have started to build a series of passages that will improve the circulation in an area with contaminated waterfalls and change their condition by building small instances of architecture that will revive the forest, which has always been a place of recreation for the people who live at its border. Banks and piers will bring back life both to the forest and to the people on the other side, who will again learn to appreciate it. This way we are creating public spaces that enhance the dignity of human space in different parts of the country.

This leads us to the following conclusion: Problems help us to enhance creativity and, above all, ethics create aesthetics. Realizing this we will gradually improve our ability to create imprints that make architecture an exercise of resolving, discovering and surprising oneself with new spaces. Thank you very much. ━━━

Wolf D. Prix

WOLF D. PRIX

My hope is that after this panel today the word sustainability will be erased from architectural language. It's the stupidest word I have heard in the last couple of years. Sustainability in German means Nach-haltigkeit. This is the ugliest word you can imagine. It's a conservative term and stems from a capatalistic topic. It was invented in the wood sector and says that you have to save property in order to get more money out of it.

Today I heard many other much better expressions for sustainabiltiy: social design, politic design, climate design and my topic today is climate design.
If we think that we save our world with small projects, that's a great misconception. I had a discussion with a famous scientist who is working in the field of climate change and he told me that we cannot avoid a climate catastrophe by building small houses with solar energy panels on the roof. We have to develop much bigger plans. The large solar plan in Africa which will be built in the next years to support the energy systems of Europe and Africa could be the first step.

One of our topics in our office is of course climate design. All our buildings consume 30% less energy than required by the codes.
It works like the example of the mid-rise apartment building which we did 15 years ago. By placing the building in the right angle to the sun and creating the wind catcher on top of the building we are using the sun to back the heating the building in winter time and the wind for cooling it in summer time.
That solar energy can also be used for cooling which was shown in the specially designed roof for the archaeological museum in Egypt.
Also our conference center in China addresses climate design issues. The façade is not only for creating energy but also designed for shading the building and directing wind so that there is a natural ventilation for the building.

The office tower we are designing for a district in Vienna is shaped by the wind because we wanted people to be able to open the windows even at a height of 120 meters. So this small measure will support the natural climate circulation and the façade we have as prototype tested right now will generate more enery than the house consumes.
That means this house is a power plant. It's conceivable that the remaining energy will be exported to the energy grid of the city. If we take the next step there will be a revolution in master planning because

Rupert Zallmann

the urban grid will no longer follow the building lines but it will actually be the energy lines, thus creating a new urban grid.

It's quite clear that at this point architecture becomes a political issue. Only when we architects can invent a new aesthetic based on the concerns of energy design will this architecture be successful. ———

TECHO EN MEXICO
THE MEXICAN ROOF
INTERVI

THE COURAGE TO RISK FAILURE

"Only a fool does not make experiments."
Charles Darwin (1809-1882)

From 1966 until 1968 Günther Feuerstein taught "Experimental Design" at the Vienna University of Technology. One of his students, Wolf D. Prix, is now continuing this legacy. Together with his students from the University of Applied Arts he not only experiments with structural engineering, aesthetics and design, but also with social patterns of society, the best example of this being the Techo in Mexico which was finished five years ago. Wojciech Czaja, the Standard's architectural journalist met the two old hands to talk about today and the past. What is the purpose of experimental design? An attempt to find answers to this question.

WOJCIECH CZAJA
According to the "Guinness Book of Records" the highest house of cards ever built consisted of 91.800 playing cards and rose to a height of 7,71 meters. It was built in Berlin in 1999. Does every experiment make sense?

——— GÜNTHER FEUERSTEIN
An experiment holds the danger or rather the possibility of failure, of collapse. This is something that is not really sought after when building in reality. There houses should and must stand for eternity. But from where can we learn how daring, how high and how lean a construction may be? Surely not from reality. Thus it makes sense to experiment just for the sake of an experiment.

CZAJA
What different kinds of experiments are there?

——— FEUERSTEIN
There is the social experiment that deals with different forms of co-existence. There is the technological experiment based on new constructions. And there is the aesthetic experiment of searching for new forms. This means that we have to clearly distinguish which kind of experiment we are talking about. My experience has been that the term "experiment" is often abused – especially in the media.

CZAJA

How important is the component of failure in an experiment?

——— *WOLF D. PRIX*

All I know is this. The "Himmelblauer" were decried as experimental architects and I deliberately say "decried". Whoever plans and builds in an experimental way does not plan and build in an orderly way, it is not stable and sound. That's what people said back then and what you still hear today. So nothing has really changed. This attitude of not allowing for any changes is, however, not so easy to detect. It is hidden behind rules and norms. A façade that has not been examined thousands of times and one that you cannot make any universally valid statements – how it will respond over the next ten years and if there is no guarantee for something then, obviously, it will never be able to prevail on the market. The problem with this is only that without experiments, attempts, there is no further development. And to put it even more bluntly: Without risking failure there is bound to be stagnation.

CZAJA

Is there any kind of framework in which this failure can be practiced and risked without hesitation?

——— *PRIX*

In experimenting with conceptual instructions each single student has the right to fail.

CZAJA

In 1966 you held the class "Experimental Design" for the first time at the Vienna University of Technology. What was your basic motivation?

——— *FEUERSTEIN*

Different thoughts and observations led up to this. At that time I had the feeling that architecture as a craft was being driven more and more into the background. Gropius had already said some decades before that architecture should go back to being a craft, back to its roots.
But nothing happened. Because of this I wanted to introduce a course where it was possible to work physically and build things with your own hands. Karl Schwanzer gave me full support on this. He also thought that each student should have hold a brick in his hand at least once during the course of his study. We finally found an empty, unused part of the Arsenal and soon the first course was held there.

—— PRIX

Here I have to add something, because this notion of experimental design did not appear out of the blue. Feuerstein wad already giving lectures on contemporary architecture at the time. These were the legendary lectures in hall 14a – and in it he had always closed the programmatic gaps of the Vienna University of Technology. Here he propagated the free form, on the other hand he recited the manifestos of incident architecture. He spoke not only about the known and established architecture of the 1950s and 1960s, but also about Mexican Pueblo architecture, Hundertwasser's Mold Manifesto, Rudolf Rudofsky's Architecture without Architects, Bruce Goff's prairie houses as well as Soleri and Archigram, etc.

CZAJA

How easy or how hard was it to integrate such a strikingly new format in the curriculum?

—— FEUERSTEIN

Actually it was quite easy. The course was introduced at the Institute of Building Theory. Schwanzer gave us his approval and you also got some credits for the course. Many students applied so there obviously was interest for the course.

CZAJA

Did you as a student profit from the course?

—— PRIX

I think the idea of taking responsibility for one's thoughts is essential. Standing on the building site with your boots in the dirt gives you completely new perspectives. Also on building.

—— FEUERSTEIN

I can remember that these courses were very popular with the students. Apart from all the technical and constructive experiments one of this course's main advantage is cooperative work. As far as I can remember this was the very first time at all at a school of higher learning such as the Vienna University of Technology that students were given the possibility to work together in a group. And all this without pencil and Aquafix drawing paper, but only with a moment of spontaneous inspiration.

CZAJA

To what extent did the student's experimental designs differ from the conventional contemporary projects?

——— *FEUERSTEIN*

They differed from the conventional project insofar as there was no design at all! For no project, for no planning or building step was there ever even a single drawing made. Nothing and never. In the best of cases we drew lines in the sand with a stick, but that was about it. We were the prototype of the paperless office. Nobody spoke about this at the time but it worked. Today everybody is talking about it, but it doesn't work at all.

CZAJA

How exactly can I envision planless building on site?

——— *FEUERSTEIN*

Karl Schwanzer was a committed and successful networker. It started with procuring building material which was then dumped at the Arsenal. There we stood in front of a pile of logs, with all the works. Mounting material, screws, clamps and nails. And so we started to build a tower from those logs.

——— *PRIX*

Ah, the tower! That was something for beginners. Helmut Swiczinsky and I soon left this stage behind us and started to work on the really interesting building task – the concrete shell. Looking back it is quite fascinating to see how quickly you can appropriate a building site and construction by a simple mark on the ground. Really atavistic!

——— *FEUERSTEIN*

The concrete shell was great, because the project soon took on a life of its own. Actually it should have been a dome. But the concrete's own weight deformed the reinforcing bars and instead of a dome we suddenly had a shell. Now we could ask ourselves whether that was already a failure. But I don't think so. By trying one thing you arrive at something different by detour.

CZAJA
If the process was really so spontaneous, who made the decisions in the case of doubt? Was there a certain person who was in command or was every step discussed in terms of grass-roots democracy?

———FEUERSTEIN
Of course we had discussions. But the foreman always had the final word. And the foreman was me.

———PRIX
Feuerstein was not a foreman, he was a manipulator. But we were also a bit crazy and ultimately simply did what we wanted to.

CZAJA
How did you select your students? Who was allowed to participate in experimental design?

———PRIX
At that time Karl Schwanzer organized the famous "Club Seminar". It was an elitist club which you could get into only by presenting especially good drafts.

———FEUERSTEIN
From the Club Seminar you could directly go to experimental design. The students from Schwanzer's Club Seminar and those in my exercises were often the same.

CZAJA
How long did that kind of a project last? And how long did the objects stand on average after completion?

———FEUERSTEIN
An amusing question, because I cannot remember witnessing any of the objects ever being removed or destroyed. But at some point and in some way it must have happened, because otherwise they would still be standing today.

———PRIX
I remember a particularly intense month. The great thing about this building site was that part of the Arsenal was marked off for this purpose. It was a playground of experimentation.

CZAJA
And then?

———— *FEUERSTEIN*
Of course even back then experimenting just for experiment's sake was not possible. We were infected by Actionism and tried to reach a certain public and to also publish our projects. I can remember that the wooden and bamboo Buckminster-Fuller dome was the project which got the most visibility and which was also featured most frequently in publications.

———— *PRIX*
I always laughed about this dome. It had already been invented and was thus neither new nor innovative. A really bad project!

———— *FEUERSTEIN*
In any case the Fuller dome gave way to the much published Fuller egg, again due to the material's own weight. Afterwards we carried the egg in a dramatic procession, which was also recorded, onto the Karlsplatz. The police didn't like that at all and disposed of it without mercy.

CZAJA
To sum up with Karl Schwanzer's words: If it is so important that students take bricks into their hands during their studies, why to this very day is this practical component not a fixed part of any curriculum?

———— *PRIX*
In the meantime the building system has become so advanced and so complex that there are separate competences for each individual step, that is, specially defined professional groups. Given the development of digital programs I also suspect that there will soon no longer be any plans at all. The trend is clearly digitalization and systematization.
As far as I know there are several schools that engage in architectural development aid. I see this as being romantic. You have students constructing schools or kindergartens on a "field trip to Africa". But this is not experimental design.

———— *FEUERSTEIN*
Craftsmanship and experimental design can only be small episodes during study. I would say that creating a separate institute or even a separate field of study of this subject is exaggerated. Architecture is an

interdisciplinary matter with many important topics. It is good that this aspect is covered but that's about it.

—— *PRIX*

But it is important that in just a few weeks you have a complete building standing in front of you! You have to keep in mind how long it takes an architect until he may finally implement his own ideas in a construction. Taking all this into consideration building at a 1:1 scale back then was an extremely important experience.

CZAJA

As you have already mentioned before, there is a big difference in planning between now and then. Experimental design at the Vienna University of Technology was planless and spontaneous, experimental design at the Angewandte on the other hand is based on concepts and careful upfront planning.

—— *PRIX*

Yes and no. That's not quite right. The roof we built five years ago in Mexico was not possible without some preliminary planning, because we had to procure materials ahead of time. The ITT, where the "Tortilla Roof" stands is in the middle of nowhere.
Purchasing bamboo later has been extremely complicated. Since we wanted to avoid tedious debates we had to plan the roof in advance so far as to be able to measure the method of construction and its dimensions. Without certain preparation it's impossible to undertake such a journey and become involved in this kind of a project.

CZAJA

What was included in the plans?

—— *PRIX*

Length, width, height. That was it. No details, no intersections, no clue.

CZAJA

Your book "The Mexican Roof" nevertheless contains very detailed plans of the object.

—— *PRIX*

These plans you are talking about were made after the project's completion for publication. As I said before, the students did not have anything to do with them except for a layout drawing with rough measurements of the roof.

CZAJA

Another difference between experimental design now and then lies in the object's functionality. While past objects where built for their own sake, the use of the Tortilla Roof in Mexico goes far beyond that.

——— PRIX

Of course. You cannot compare Arsenal with Mexico. The situation is totally different. It is not only about constructive experimentation but also about social usage. After all the roof should become a jump start for a whole training centre on the hills. But on the other hand that's the big danger with such a project. You have to be terribly careful that it does not turn into some sort of development aid or architectural colonization. Suddenly students from the 1st World break into the 3rd World as though they were some conquistadores from Spain or Portugal and build their own view of contemporary architecture in the local landscape.

CZAJA

What else must be taken into consideration when experimental architecture suddenly sports a function?

——— PRIX

My observation has been that with each additional function the project's iconicity is gradually lost. The more something has to achieve, the more common and ordinary it gets. Then you suddenly arrive at a point where you cannot detect the originally intended symbol or icon anymore.

CZAJA

Is experimental design then about conveying something or about the experiment?

——— PRIX

Each experiment should also signal something. Emblematic forms provide a means of identification.

CZAJA

Is the experiment still in the foreground when all technical possibilities are exhausted? Through CAD programs and 3D modeling a large part of experimenting has been transferred to virtual space. What influence has this technological development on the true building experiment on site?

———PRIX

The influence of computer language on architecture is clearly visible today. The machine has more potential than the pencil. I could never draw a building deformed by wind, a building in which you can open the window even at a height of 200 m. The computer, though, is not spontaneous, not creative and it is not flexible. It gives us certain possibilities but it cannot assume responsibility. This is something that we ourselves have to take on.

———FEUERSTEIN

A computer is a splendid tool for formal and aesthetic experiments. There are great possibilities arising from the formal richness of the medium. To be honest: You will never achieve this with a group of students working on site. But the great weakness of CAD is that it cannot connect constructive, aesthetic and social parameters.

———PRIX

There is one thing I have to add. The computer is a great instrument for testing decisions already made and final concepts. Where else would I be able to see how the building loses its form under great stress or in the case of an earthquake? You type this into the machine and a few hours later it spits out the answer.

CZAJA

Mr. Feuerstein, in your book "Visionäre Architektur Wien 1958/1988" you write explicitly that back then the process was more important than the result. Is this still true?

———FEUERSTEIN

It must be assessed differently from a pedagogical point of view than, for example, from an architectural point of view. For me personally the sociopolitical and social component of experimental design was very important. Maybe I took this factor too seriously. Maybe that was a kind of flaw in my lecture. I cannot say for sure.

CZAJA

And what is more important today: process or result?

PRIX

Process and result cannot be separated.

CZAJA

In a project like the Roof in Mexico, I assume, the result is much more important than the process behind it. After all, it was about constructing a functioning building.

PRIX

Again: Those two things cannot be looked at separately.

CZAJA

Would your students have been happy, would the Mexicans have been happy if experimental design in this particular project had failed?

PRIX

What's failure? If we hadn't built the roof or if it had been destroyed by a hurricane? I would say that as an architect you fail because you are runner-up in a competition. And true failure is if a student is extremely gifted and you never hear from him again.

CZAJA

Mr. Prix, Günther Feuerstein in one of his articles about "Experimental Design" called you an extremely committed and high-profiled student. Is this what you're trying to get at?

PRIX

No, of course not. I am talking in general terms. But I also ask myself if the word "failure" really has the negative connotation as we are just claiming.

FEUERSTEIN

Of course it's negative! What is positive about failure?

PRIX

And if somebody said: "Leonardo da Vinci failed with his aircrafts." Is that a legitimate statement? Because it is true that Leonardo was way ahead of his time. He didn't build an aircraft but he formulated the idea of a flying human.

——— FEUERSTEIN

Could he fly these aircrafts as he originally intended? No, he could not. So he failed.

——— PRIX

Yes, right. But the idea and the drawings are beautiful!

CZAJA

Back to Mexico. The Tortilla Roof's construction was at one time so laden that it could not bear any more weight. Speedily the guardian angel Franz Sam – there you are - was flown in and aided with statics and construction.

——— PRIX

Even if the roof had collapsed I would only have described it as a misfortune, but not a failure. Isn't failure a part of experimenting?

CZAJA

What is the difference between failure and a misfortune?

——— PRIX

Czaja! That's stuff for the next interview.

——— FEUERSTEIN

Hold on! If a building collapses you have to admit to yourself that the project is definitely a failure. The roof of the Gasometer's concert hall collapsed, a sports hall in Bavaria collapsed, the Reichsbrücke in Vienna collapsed….and there are enough other examples. What else should you call it? And what a failure it is! On the other hand you have to distinguish in what sector failure happens.

CZAJA

An example?

——— FEUERSTEIN

Take a look for example at the Stadt des Kindes (City of Children) by Anton Schweighofer: It is an innovative and trend-setting project, which is a building specifically geared to children. In architecture Schweighofer accomplished something really big, but he failed in the social experiment of overcoming isolation on the city's periphery.

CZAJA

We have now talked about the past and the future. How will experimental design change in the future? In what direction will experiment and risk go?

———*PRIX*

It will definitely continue. Feuerstein and I have experimented with concrete, bamboo and wood. The next step is going to be building with prestressed adobe, a clay building material which cannot only absorb pressure but also pulling force. At the moment we are preparing a project for Mexico. In two or three years I guess we will be ready to create a building with a prestressed adobe construction. That could be a breakthrough in Pueblo architecture.

CZAJA

And you?

———*FEUERSTEIN*

No more experiments. My private and professional experiments are long, past and over. Was it an experiment that I married a woman 40 years younger than myself? Some say yes. I don't know. This experiment has in any case been quite successful until now.

———*PRIX*

Private experiments are the best ones anyway! That's like driving into a turn with 200 which only takes 90 km/h. That might be dangerous, but it is extremely exciting.

CZAJA

Many people had to experience this unpleasant failure.

———*PRIX*

Yes, it is the ultimate failure. I am probably intelligent enough to slow down to 150.

CZAJA

Shall we end with this topic?

———*PRIX*

Of course not – we think positively, right. I envision an experiment that could change architecture and revolutionize building. But this vision is no longer a dream as the experiment is already being tested. I'm thinking here of the BIM (Building Information Modeling) computer program which could give the architect his status again, which he has lost, namely that he can decide over his building again. This method which we and other architects are testing enables us to capture and precisely define details already in the design stage.

This will quell all unnecessary squabbles about the so-called project management in the future. And the architect will once again be a partner of the client and not simply a service provider. Only this way can architecture move forward and advance.

——— *FEUERSTEIN*
That's a risky experiment.

——— *PRIX*
Of course, and I hope it won't fail. However, my impression is that the training of architects at mass universities is not moving in this direction. And that the young architects of the future are being forced to bow to authority. This simply cannot continue.

CZAJA
That's an equally awful closing remark.

——— *PRIX*
You simply have to say it as is. ———

TECHO EN MEXICO
THE MEXICAN ROOF
TEAM—

Rüdiger Suppin

Alexander Matl

Andrea Börner

Franz Sam

Jean Pierre Bolivar Martinez

Giulio Polita

Rupert Zallmann

Bärbel Müller

Florian Schafschetzy

TECHO EN MEXICO
THE MEXICAN ROOF REVISITED 2009

Editor: Wolf D. Prix
Editors in Chief: Bärbel Müller, Roswitha Janowski-Fritsch
Design: Christian Sulzenbacher
Photo Credits: Andrea Börner, Pedro Cortés,
Bärbel Müller, Franz Sam, Peter Strasser, Reiner Zettl
Image Editing: Andreas Wenk
Translation: Wolfgang Dallasera, Camilla Nielsen,
Barbara Strohmaier, Richard Watts
Printing: Holzhausen Druck GmbH

© 2011 Springer-Verlag/Wien
Printed in Austria
SpringerWienNewYork is a part of
Springer Science + Business Media
springer.at

Printed on acid-free and chlorine-free bleached paper

SPIN: 80026738

Library of Congress Control Number: 2010939639

ISSN 1866-248X
ISBN 978-3-7091-0471-2
SpringerWienNewYork

 INSTITUTE OF ARCHITECTURE